ANIMAL GROUPS

A RAFT OF OTTERS

BY SETH LYNCH

Gareth Stevens PUBLISHING

Please visit our website, www.garethstevens.com. For a free color catalog of all our high-quality books, call toll free 1-800-542-2595 or fax 1-877-542-2596.

Library of Congress Cataloging-in-Publication Data

Names: Lynch, Seth, author.
Title: A raft of otters / Seth Lynch.
Description: New York : Gareth Stevens Publishing, [2022] | Series: Animal groups | Includes index. | Audience: Ages 6 | Audience: Grades K-1
Identifiers: LCCN 2020028904 (print) | LCCN 2020028905 (ebook) | ISBN 9781538263556 (library binding) | ISBN 9781538263532 (paperback) | ISBN 9781538263549 (set) | ISBN 9781538263563 (ebook)
Subjects: LCSH: Otters–Juvenile literature. | Otters–Behavior–Juvenile literature.
Classification: LCC QL737.C25 L954 2022 (print) | LCC QL737.C25 (ebook) | DDC 599.769–dc23
LC record available at https://lccn.loc.gov/2020028904
LC ebook record available at https://lccn.loc.gov/2020028905

Published in 2022 by
Gareth Stevens Publishing
111 East 14th Street, Suite 349
New York, NY 10003

Designer: Tanya Dellaccio
Editor: Kristen Nelson

Photo credits: Cover Hal Beral/Corbis/Getty Images; p. 5 Terence/Shutterstock.com; p. 7 Santiparp Wattanaporn/Shutterstock.com; p. 9 superjoseph/Shutterstock.com; p. 11 Neil Aronson/Shutterstock.com; pp. 13, 17 MODpix/Shutterstock.com; p. 15 worldswildlifewonders/Shutterstock.com; p. 19 Michael0706/Shutterstock.com; p. 21 Doug Meek/Shutterstock.com.

Printed in the United States of America

CPSIA compliance information: Batch #CSGS22: For further information contact Gareth Stevens, New York, New York at 1-800-542-2595.

CONTENTS

Boldface words appear in the glossary.

Fun with Otters

Did you know there are about 13 kinds of otters? All of them can be playful and **social** animals, but it's common for otters to mostly live alone. Sea otters, though, live in groups in their ocean home. Let's learn more about sea otters!

From Fur to Feet

Sea otters are **mammals**. Their bodies are covered in gray or brown fur. This fur is thick and waterproof. It keeps them warm in the water. Sea otters also have webbed feet that help them swim well.

Life at Sea

Sea otters live in the Pacific Ocean off the coast of North America and Asia. Most live off the coast of the U.S. state of Alaska. They may leave the ocean at times, but they mostly spend their lives in the water.

Getting Together

Scientists have seen as many as 2,000 otters resting together in the water! When otters float on their backs together in a group, it's called a raft. They use a seaweed called kelp to keep them from floating away from each other.

A group of otters may also be called a pod. Some pods are made up of fewer than 10 otters. Other pods may have hundreds of otters! Sea otter pods are commonly either all male or made up of females and young.

Male and female groups mix when they look for food. They also meet when it's time to **mate**. After about six months, mother otters give birth in the water. They're the only kind of otter to do that!

Sea Otter Pups

A mother sea otter gives birth to just one baby, or pup, at a time. She floats on her back to hold and feed her pup. Mothers and pups stay together for many months. Mothers teach pups to swim and hunt for food.

Fishy Meal

Sea otters eat while floating on their backs too! They catch their **prey** by diving to the bottom of their ocean **habitat**. They use rocks to open the shells of clams, sea urchins, and crabs. Sea otters also eat fish and octopus.

In Danger

Sea otters used to be hunted for their fur. This made their **population** much smaller. Sea otters are now **protected** from hunting by laws. But **pollution** and other human activities still harm them. There aren't many left in the wild. It's our job to keep sea otters safe!

GLOSSARY

habitat: the natural place where an animal or plant lives

mammal: a warm-blooded animal that has a backbone and hair, breathes air, and feeds milk to its young

mate: to come together to make babies

pollution: matter that makes land, water, or air dirty or not safe for people and animals

population: the number of animals of the same kind that live in a place

prey: an animal that is hunted by other animals for food

protect: to keep safe

social: having to do with liking being with others

FOR MORE INFORMATION

BOOKS

Felix, Rebecca. *Otterly Adorable Otters*. Minneapolis, MN: Lerner Publications, 2021.

Suen, Anastasia. *Sea Otters*. Mankato, MN: Amicus/Amicus Ink, 2020.

WEBSITES

Sea Otters
kids.nationalgeographic.com/animals/mammals/sea-otter/
See many pictures of cute otters on this website.

Sea Otters
www.montereybayaquarium.org/animals/animals-a-to-z/sea-otter
Find out more about sea otters here!

INDEX